EVANSTON PUBLIC LIBRARY
S0-BNW-234

MathStart
MATCHING

A Pair of Socks

by Stuart J. Murphy

illustrated by Lois Ehlert

HarperCollinsPublishers

To Kristin—whose great ideas outnumber her lost socks
—S.J.M.

The collage illustrations in this book were created with Pantone coated papers, cut with scissors, and glued to Strathmore paper.

For more information about the MathStart series, please write to HarperCollins Children's Books, 10 East 53rd Street, New York, NY 10022.

Bugs incorporated in the MathStart series design were painted by Jon Buller.

A Pair of Socks

Library of Congress Cataloging-in-Publication Data
Murphy, Stuart J., date
A pair of socks / by Stuart J. Murphy ; illustrated by Lois Ehlert.
p. cm. (MathStart. Level 1)
Summary: Introduces pattern recognition as a sock searches the house for its lost mate.
ISBN 0-06-025879-9. — ISBN 0-06-025880-2 (lib. bdg.)
ISBN 0-06-446703-1 (pbk.)
1. Pattern perception—Juvenile literature. [1. Pattern perception.]
I. Ehlert, Lois, ill. II. Title. III. Series.
BF293.M87 1996 95-19618
153.14'23—dc20 CIP
AC

Typography byTom Starace
1 2 3 4 5 6 7 8 9 10
❖
First Edition

A Pair of Socks

I'll never be worn.
It doesn't seem fair.

I'm missing my match—
I'm not part of a pair.

LAUN

This one's stinky and grimy–

and not quite like me.

This one's
sudsy and
slimy–

not the same,
I can see.

This one's
all warm
and fluffy—

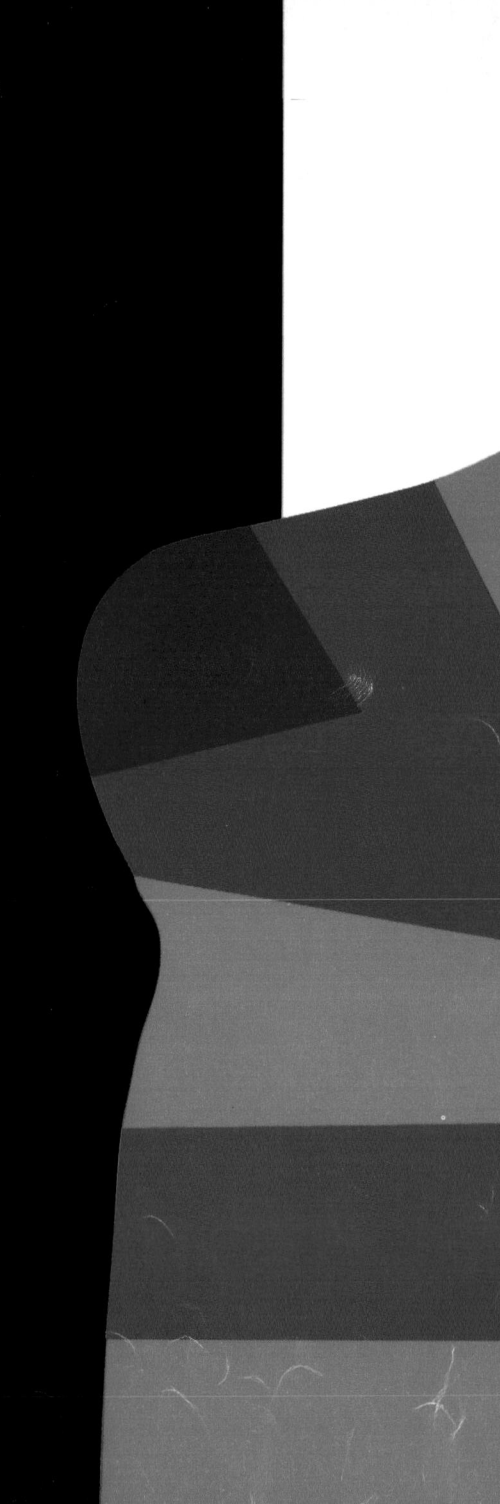

not just
red and blue.

This one's folded and puffy–

but spots are
wrong, too.

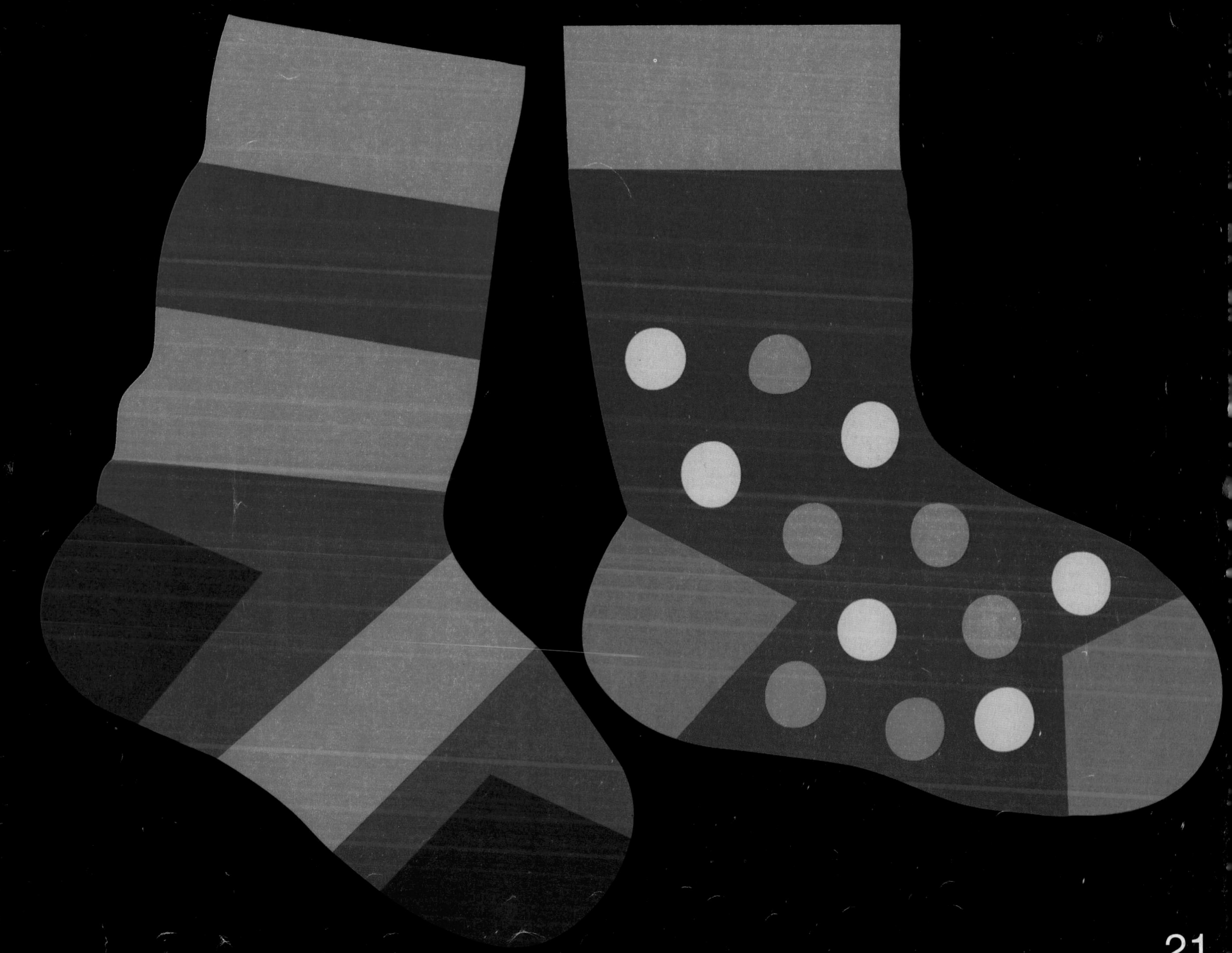

I've been grabbed by the pup!

His basket's
not far. . . .

I was going to give up, but *here* you are!

I finally found you,
but my heel has been torn.
What terrible luck:
We still won’t be worn.

Our problem is solved with a simple blue patch.

We'll travel together, me and my match.

Can you find the match for each sock?

FOR ADULTS AND KIDS

If you would like to have more fun with the math concepts presented in *A Pair of Socks*, here are a few suggestions:

- Read the story together and ask the child to describe what is going on in each picture.
- Ask questions throughout the story, such as "Are the socks the same?" "How is one sock different from the other sock?" and "Which is your favorite sock?"
- Together draw and color some pairs of socks in a variety of patterns. Then cut them out and separate the pairs. Play a game of matching the socks.
- Gather some matched and mismatched household items—such as mittens, socks or shoes, napkins, place mats, or towels—and talk about them together using the vocabulary from the book. For example: "Which shoes are the same?" "Which towels are different?" "How are they different?"
- Look at things in the real world—wallpaper, rugs, floor tiles, etc.—and discuss their patterns. Draw pictures of the patterns that you find. Make up your own patterns.

Following are some activities that will help you extend the concepts presented in *A Pair of Socks* into a child's everyday life.

Cooking: Decorate cookies or cupcakes with different colors of icing, sprinkles, or candies. Then arrange them on plates in patterns, such as red-red-green-red-red-green, etc. Ask: "Which pattern do you like best?"

Nature: Collect a handful of leaves and pile them up in two or three different size categories. Arrange the leaves in patterns, such as big-medium-small-small-big-medium-small-small, etc. Change the patterns.

Music: Play follow-the-leader with different clapping and stomping patterns, such as clap-clap-stomp-clap-clap-stomp, etc.

Crafts: Collect buttons of different colors and sizes and, using string or yarn, make necklaces that contain patterns.

The following books include some of the same concepts that are presented in *A Pair of Socks*:

- DOTS, SPOTS, SPECKLES, AND STRIPES by Tana Hoban
- KNOWABOUT PATTERN by Henry Pluckrose
- CAPS FOR SALE by Esphyr Slobodkina